Monism

~ AS CONNECTING ~

Religion AND Science

THE CONFESSION OF FAITH BY A MAN OF SCIENCE

~ *WRITTEN BY* ~

Ernst Heinrich Philipp August Haeckel

~ TRANSLATRED FROM GERMAN BY ~

James Grant Gilchrist

~ EDITED AND FOREWORD BY ~

Orry William von Diez

First Published in Berlin, German Empire, 1892

First English Publication in London, British Empire, 1894

First Publication of this Edition in the American Empire, 2024

ISBN

~

978-1-7348567-3-6

Page Listing

Editors Foreword

Born Ernst Heinrich Philipp August Haeckel on the 16th February, 1834 in Potsdam, Brandenburg, Kingdom of Prussia. Studying medicine at universities in Berlin and Würzburg, he earned his doctorate in 1857. Engaging professionally in many fields including zoology, botany, philosophy, human medicine and even art, Haeckel was a man of many talents and interests.

Doctor Haeckel, to put it lightly, was a genius of a scientist and one the most important men of Academia in Germany during the latter half of the 19th century. Indeed, Haeckel was so monumental in his efforts to promote Darwinian Evolution, that it is said that he is more responsible for the popularity of Darwin on the European continent than is Darwin himself. Known most commonly for his vibrant and intricately detailed artistic works of his various research subjects, Ernst was most enamoured by marine biology. His most widely circulated work is a collection of artistic pieces entitled *Kunstformen der Natur (Art Forms of Nature),* which heavily focuses on various species of marine life. He was personally responsible for the discovery and classification of quite a few previously-unknown species across several biomes, adding a plethora of information to the ever expanding catalogue of natural wonders Haeckel was so meticulous in noting.

He was also an avowed Naturalist and proponent of the burgeoning ideas of eugenics in his era. Haeckels work influenced a long line of thinking in regards to Naturalism, Human Evolution, and Social theories. One of the most important connections to have been made, was the research conducted by Dr. Wolfgang Köhler (1887-1967) regarding Gestalt[1] Psychology. Dr. Köhler was the head of the Anthropoid Research Station of the Prussian Academy of Sciences on the island of Tenerife in the Canary Islands (off the Northwest coast of Africa) in 1913. Here he conducted research on simians, most notably chimpanzees. Köhler had not seemingly studied directly under Haeckel, but was nonetheless heavily influenced by Haeckels already prolific contributions to many fields Köhler was engaged with. In 1917, Köhler published one of his most influential books, *The Mentality of Apes* (coupled with the German *Intelligenzprüfungen an Anthropoiden* [*Intelligence Tests on Anthropoids*]). It was this type of cross-species research in evolutionary studies that solidified the ideas of Gestalt for Köhler, and prior to his death in 1919, Haeckel aswell. There is a permeating notion of *Oneness* that is conceptualized by the understanding of Gestalt; inthat this all-encompassing totality of matter and non-matter, being and un-being, that which is all *things-unto-themselves*[2], all that which is *a-priori* and *a-posteriori*[3]– simply that '*it*' *is*, and that *it* can be called Nature, God, the Cosmos, or given to the Sisyphusian task of achieving GodHead; but only that *it is total* in its form, and that all things in it either strive for concert or chaos in their workings of and within the *whole*. Monism, as supported by Haeckel, is very much in line with Gestalt principles of human and animal thought. This Continuity of thought, spanning many thousands of years of human history, expresses a desire to understand the very real connection we all share to and exist within Nature.

The opening of this particular essay has a tendency to catch one off guard, as Haeckel initiates the presentation rapidly and aggressively. He is seen defending and elucidating, briefly in speech format, his and his communities ideas of Monistic philosophy and how these views reconcile Science and Spirituality. And they do not cease at mere reconciliation, on the contrary, Haeckel sought to prove inseparable the realms of physics and metaphysics, those of the Corporeal and the Spiritual. Firmly holding true to the Laws of Nature and objective and organic hierarchy, Haeckel was a proponent of Naturalism and subsequently also of Eugenics; believing that evolutionary principles of selective breeding and culling could be applied to humans, as is done across the Animal Kingdom, in order to refine the various human breeds and species to their respective benefits. He believed that eugenics was a useful medical tool in preventing the spread of debilitating and terminal illnesses, and that with enough directed genetic hygienics we could eliminate unnecessary hereditary suffering altogether from whole populations, sparing future generations of needless misery.

In our decidedly modern, yet wholly barbaric, theatre of Western Academia, we find ourselves ever more pushed to the brink of total ignorance. If the previous so-called dark ages over a millennia ago were considered so, insofar that they were inhabited by a population devoid of understanding and reason onset by a lack of information; then this modern dark age, picking up nearly a thousand years after the last left off, brings us a lack of reason and understanding from an over-abundance of nonsensical information. People in the modern dark age have no idea what to believe, who to believe, or what is real. And this confusion is brought upon by the sorcery of the age; media of all types, be it television, radio, video games, books, comics etc. etc. *ad nausea*. A constant stream of information, good, bad and utterly useless, bombards the individual to the point of being overwhelmed. Haeckel's work has been sidelined by mainstream academia, and it is precisely because his, and similar, works cut straight to the bone and deliver sense in a senseless world.

Only through understanding the principles of Nature itself, only through understanding our place in the universe and its workings, can we finally be free of inner chaos and experience inner harmony, such that we can overcome our animal-selves and burst forth into the stars. Endless struggle awaits us beyond the veil of our humble skies, and it is in that struggle that the pressures of evolution shall truly shape the fate of Man.

O. W. von Diez

Hanover, Pennsylvania, Empire of the United States, 2024

[1] [*study of all things in relation to all things as separate but symbiotic parts to a unifying greater whole comprised of all things – to study something for a holistic and total perspective*]

[2] [see: Immanuel Kant & Arthur Schopenhauer - *Ding(e) an sich selbst* (*Thing in Itself, Things unto themselves*)]

[3] *a-priori [a truth that is by its own reason knowable to naked instincts of the mind] [a truth that is knowable sans exterior stimuli]*

a-posteriori [a truth that is made evident and/or knowable only after having interacted with elements of it which then present themselves as truth] [a truth that is knowable post exterior stimuli]

Authors Preface

The following lecture on Monism is an informal address delivered extemporaneously on October 9, 1892, at Altenburg, on the seventy-fifth anniversary of the "Naturforschende Gesellschaft des Osterlandes." The immediate occasion of it was a previous address delivered by Professor Schlesinger of Vienna on "Scientific Articles of Faith." This philosophical discourse contained, with reference to the weightiest and most important problems of scientific investigation, much that was indisputable; but it also contained some assertions that challenged immediate rejoinder and a statement of the opposite view. As I had for thirty years been very closely occupied with these problems of the philosophy of nature, and had set forth my convictions with respect to them in a number of writings, a wish was expressed by several members of the Congress that on this occasion I should give a summary account of these. It was in compliance with this wish that the following "Scientific Confession of Faith" was uttered. The substance of it, as written from recollection on the day after its delivery, first appeared in the *Altenburger Zeitung of* 19th October 1892. This was reproduced, with one or two philosophical additions, in the November number *of* the *Freie Bühne für den Entwickelungskampf der Zeit* (Berlin). In its present form the Altenburg address is considerably enlarged, and some parts have been more fully worked out. In the notes [pg.]* several burning questions of the present day *have* been dealt with from the monistic point of view.

*[*Originally this was cited as page 91 in the original 1894 Gilchrist translation.*]

The purpose of this candid confession of monistic faith is twofold. First, it is my desire to give expression to that rational view of the world which is being forced upon us with such logical rigour by the modern advancements in our knowledge of nature as a unity, a view in reality held by almost all unprejudiced and thinking men of science, although but few have the courage (or the need) to declare it openly. Secondly, I would fain establish thereby a bond between religion and science, and thus contribute to the adjustment of the antithesis so needlessly maintained between these, the two highest spheres in which the mind of man can exercise itself; in monism the ethical demands of the soul are satisfied, as well as the logical necessities of the understanding.

The rising flood of pamphlets and books published on this subject, demonstrates that such a natural union of faith and knowledge, such a reasonable reconciliation of the feelings and the reason, are daily becoming a more pressing necessity for the educated classes. In North America (in Chicago), there has been published for several years a weekly journal devoted to this purpose: *The Open Court: A Weekly Journal devoted to the Work of Conciliating Religion and Science*. Its worthy editor, Dr. Paul Carus (author of *The Soul of Man*, 1891), devotes also to the same task a quarterly journal under the title *The Monist*. It is in the highest degree desirable that so worthy endeavours to draw together the empirical and speculative views of nature, realism and idealism, should have more attention and encouragement than they have hitherto received, for it is only through a natural union of the two that we can approach a realization of the highest aim of mental activity-the blending of religion and science in monism.

Ernst Haeckel

Jena, Grand Duchy of Saxe-Weimar-Eisenbach
(Thüringian States), German Empire, 1892

Ernst Haeckel.
Geboren 16. Februar 1834 in Potsdam.

Monism as Connecting Religion and Science

by

Ernst H. P. A. Haeckel

XIX

A society for investigating nature and ascertaining truth cannot celebrate its commemoration day more fittingly than by a discussion of its highest general problems. It must be regarded, therefore, with satisfaction that the speaker on such an august occasion as this—the seventy-fifth anniversary of your Society—has selected as the subject of his address a theme of the highest general importance. Unfortunately, it is becoming more and more the custom on such occasions, and even at the general meetings of the great "Association of German Naturalists and Physicians," to take the subject of address from a narrow and specialized territory of restricted interest. If this growing custom is to be excused on the grounds of increasing division of labour and of diverging specialization in all departments of work, it becomes all the more necessary that, on such anniversaries as the present, the attention of the audience should be invited to larger matters of common interest.

Such a topic, supreme in its importance, is that concerning "Scientific Articles of Faith," upon which Professor Schlesinger has already expounded his views.[1] I am glad to be able to agree with him in many important points, but as to others I should like to express some hesitation, and to ask consideration for some views which do not coincide with his. At the outset, I am entirely at one with him as to that unifying conception of nature as a whole which we designate in a single word as Monism. By this we unambiguously express our conviction that there lives "one spirit in all things," and that the whole cognisable world is constituted, and has been developed, in accordance with one common fundamental law. We emphasize by it, in particular, the essential unity of inorganic and organic nature, the latter having been evolved from the former only at a relatively late period.[2] We cannot draw a sharp line of distinction between these two great divisions of nature, any more than we can recognize an absolute distinction between the animal and the vegetable kingdom, or between the lower animals and man. Similarly, we regard the whole of human knowledge as a structural unity; in this sphere we refuse to accept the distinction usually drawn between the natural and the spiritual. The latter is only a part of the former (or *vice versa*); both are one. Our monistic view of the world belongs, therefore, to that group of philosophical systems which from other points of view have been designated also as mechanical or as pantheistic. However differently expressed in the philosophical systems of an Empedocles or a Lucretius, a Spinoza or a Giordano Bruno, a Lamarck or a David Strauss, the fundamental thought common to them all is ever that of the oneness of the cosmos, of the indissoluble connection between energy and matter, between mind and embodiment—or, as we may also say, between God and the world—to which Goethe, Germany's greatest poet and thinker, has given poetical expression in his *Faust* and in the wonderful series of poems entitled *Gott und Welt*.

That we may rightly appreciate what this Monism is, let us now, from a philosophico-historical point of view cast a comprehensive glance over the development in time of man's knowledge of nature. A long series of varied conceptions and stages of human culture here passes before our mental vision. At the lowest stage, the rude—we may say animal—phase of prehistoric primitive man, is the "ape-man," who, in the course of the tertiary period, has only to a limited degree raised himself above his immediate pithecoid ancestors, the anthropoid apes. Next come successive stages of the lowest and simplest kind of culture, such as only the rudest of still existing primitive peoples enable us in some measure to conceive. These "savages" are succeeded by peoples of a low civilization, and from these again, by a long series of intermediate steps, we rise little by little to the more highly civilized nations. To these alone—of the twelve races of mankind only to the Mediterranean and Mongolian—are we indebted for what is usually called "universal history." This last, extending over somewhat less than six thousand years, represents a period of infinitesimal duration in the long millions of years of the organic world's development.

Neither of the primitive men we have spoken of, nor of those who immediately succeeded them, can we rightly predicate any knowledge of nature. The rude primitive child of nature at this lowest stage of development is as yet far from being the restless *Ursachenthier* (origin-seeking animal) of Lichtenberg; his demand for causes has not yet risen above that of apes and dogs; his curiosity has not yet mounted to pure desire of knowledge. If we must speak of "reason" in connection with pithecoid primitive man, it can only be in the same sense as that in which we use the expression with reference to those other most highly developed Mammals, and the same remark holds true of the first beginnings of religion.[3]

It is indeed still not infrequently the custom to deny absolutely to the lower animals reason and religion. An unprejudiced comparison, however, convinces us that this is wrong. The slow and gradual process towards completeness which, in the course of thousands of years, civilized life has been working in the soul of man, has not passed away without leaving some trace on the soul of our highest domestic animals also (above all, of dogs and horses). Constant association with man, and the steady influence of his training, have gradually, and by heredity, developed in their brain higher associations of ideas and a more perfect judgment. Drill has become instinct, an undeniable example of "the transmission of acquired characters."[4]

Comparative psychology teaches us to recognize a very long series of successive steps in the development of soul in the animal kingdom. But it is only in the most highly developed vertebrates-birds and mammals—that we discern the first beginnings of reason, the first traces of religious and ethical conduct. In them we find not only the social virtues common to all the higher socially-living animals,—neighborly love, friendship, fidelity, self-sacrifice, etc.,—but also consciousness, sense of duty, and conscience; in relation to man their lord, the same obedience, the same submissiveness, and the same craving for protection, which primitive man in his turn shows towards his "gods." But in him, as in them, there is yet wanting that higher degree of consciousness and of reason, which strives after a *knowledge* of the surrounding world, and which marks the first beginning of philosophy or "wisdom." This last is the much later attainment of civilized races; slowly and gradually has it been built up from lower religious conceptions.

At all stages of primitive religion and early philosophy, man is as yet far removed from monistic ideas. In searching out the causes of phenomena, and exercising his understanding thereon, he is in the first instance prone in every case to regard personal beings—in fact, anthropomorphic deities—as the agents at work. In thunder and lightning, in storm and earthquake, in the circling of sun and moon, in every striking meteorological and geological occurrence, he sees the direct activity of a personal god or spirit, who is usually thought of in a more or less anthropomorphic way. Gods are distinguished as good and bad, friendly and hostile, preserving and destroying, angels and devils.

This becomes true in a yet higher degree when the advancing pursuit of knowledge begins to take into consideration the more complicated phenomena of organic life also, the appearance and disappearance of plants and animals, the life and death of man. The constitution of organized life, so suggestive as it is of art and purpose, leads one at once to compare it with the deliberately designed works of man, and thus the vague conception of a personal god becomes transformed into that of a creator working according to plan. As we know, this conception of organic creation as the artistic work of an anthropomorphic god—of a divine mechanic—generally maintained its ground almost everywhere, down even to the middle of our own century, in spite of the fact that eminent thinkers had demonstrated its untenability more than two thousand years ago. The last noteworthy scientist to defend and apply this idea was Louis Agassiz (died 1873). His notable *Essay on Classification*, 1857, developed that theosophy with logical vigor, and thereby reduced it to an absurdity.[5]

All these older religious and teleological conceptions, as well as the philosophical systems (such as those of Plato and of the Church fathers) which sprang from them, are antimonistic; they stand in direct antithesis to our monistic philosophy of nature. Most of them are dualistic, regarding God and the world, creator and creature, spirit and matter, as two completely separated substances. We find this express dualism also in most of the purer church-religions, especially in the three most important forms of monotheism which the three most renowned prophets of the eastern Mediterranean—Moses, Christ, and Mohammed—founded. But soon, in a number of impure varieties of these three religions, and yet more in the lower forms of paganism, the place of this dualism is taken by a philosophical pluralism, and over against the good and world-sustaining deity (Osiris, Ormuzd, Vishnu), there is placed a wicked and destroying god (Typhon, Ahriman, Shiva). Numerous demigods or saints, good and bad, sons and daughters of the gods, are associated with these two chief deities, and take part with them in the administration and government of the cosmos.

In all these dualistic and pluralistic systems the fundamental idea is that of anthropomorphism, or the humanizing of God; man himself, as godlike (or directly descended from God), occupies a special position in the world, and is separated by a great gulf from the rest of nature. Conjoined with this, for the most part, is the anthropocentric idea, the conviction that man is the central point of the universe, the last and highest final cause of creation, and that the rest of nature was created merely for the purpose of serving man. In the Middle Ages there was associated at the same time with this last conception the geocentric idea, according to which the earth as the abode of man was taken for the fixed middle point of the universe, round which sun, moon, and stars revolve. As Copernicus (1543) gave the death-blow to the geocentric dogma, so did Darwin (1859) to the anthropocentric one closely associated with it.[6] A broad historical and critical comparison of religious and philosophical systems, as a whole, leads as a main result to the conclusion that every great advance in the direction of profounder knowledge has meant a breaking away from the traditional dualism (or pluralism) and an approach to monism. Ever more clearly are we compelled by reflection to recognize that God is not to be placed over against the material world as an external being, but must be placed as a "divine power" or "moving spirit" within the cosmos itself. Ever clearer does it become that all the wonderful phenomena of nature around us, organic as well as inorganic, are only various products of one and the same original force, various combinations of one and the same primitive matter. Ever more irresistibly is it borne in upon us that even the human soul is but an insignificant part of the all-embracing "world-soul"; just as the human body is only a small individual fraction of the great organized physical world.

*T*he great general principles of theoretical physics and chemistry are now in a position to afford to this unifying conception of nature an exact, to a certain extent, indeed, a mathematical confirmation. In establishing the law of the "conservation of energy," Robert Mayer and Helmholtz showed that the energy of the universe is a constant unchangeable magnitude; if any energy whatever seems to vanish or to come anew into play, this is only due to the transformation of one form of energy into another. In the same way Lavoisier's law of the "conservation of matter" shows us that the material of the cosmos is a constant unchangeable magnitude; if any body seems to vanish (as, for example, by burning), or to come anew into being (as, for example, by crystallization), this also is simply due to change of form or of combination. Both these great laws—in physics, the fundamental law of the conservation of energy, and in chemistry, of the conservation of matter—may be brought under one philosophical conception as the law of the conservation of substance; for, according to our monistic conception, energy and matter are inseparable, being only different inalienable manifestations of one single universal being-substance.[7] In a certain sense we can regard the conception of "animated atoms" as essentially partaking of the nature of this pure monism—a very ancient idea which more than two thousand years ago Empedocles enunciated in his doctrine of "hate and love of the elements." Modern physics and chemistry have indeed in the main accepted the atomic hypothesis first enunciated by Democritus, in so far as they regard all bodies as built up of atoms, and reduce all changes to movements of these minutest-discrete particles. All these changes, however, in organic as well as in inorganic nature, become truly intelligible to us only if we conceive these atoms not as dead masses, but as living elementary particles endowed with the power of attraction and repulsion. "Pleasure" and "pain," and "love" and "hate," as predicates of atoms are only other expressions for this power of attraction and repulsion.

Although, however, monism is on the one hand for us an indispensable and fundamental conception in science, and although, on the other hand, it strives to carry back all phenomena, without exception, to the mechanism of the atom, we must nevertheless still admit that as yet we are by no means in a position to form any satisfactory conception of the exact nature of these atoms, and their relation to the general space-filling, universal ether. Chemistry long ago succeeded in reducing all the various natural substances to combinations of a relatively small number of elements; and the most recent advances of that science have now made it in the highest degree probable that these elements or the (as yet) irreducible primitive materials are themselves in turn only different combinations of a varying number of atoms of one single original element. But in all this we have not as yet obtained any further light as to the real nature of these original atoms or their primal energies.

A number of the acutest thinkers have, so far in vain, endeavored to grapple more closely with this fundamental problem of the philosophy of nature, and to determine more exactly the nature of atoms as well as their relation to the space-filling ether. And the idea steadily gains ground that no such thing as empty space exists, and that everywhere the primitive atoms of ponderable matter or heavy "mass" are separated from each other by the homogeneous ether which extends throughout all space. This extremely light and attenuated (if not imponderable) ether causes, by its vibrations, all the phenomena of light and heat, electricity and magnetism. We can imagine it either as a continuous substance occupying the space between the mass-atoms, or as composed of separate particles; in the latter case we might perhaps attribute to these ether-atoms an inherent power of repulsion in contrast to the immanent attracting power of the heavy mass-atoms, and the whole mechanism of cosmic life would then be reducible to the attraction of the latter and the repulsion of the former. We might also place the "vibrations of the cosmic ether" alongside of the "operation of space in general," in the sense in which these words are used by Professor Schlesinger.

At any rate, theoretical physics has in recent years made an advance of fundamental importance and widest reach in our knowledge of nature, in that it has come nearer to a knowledge of this cosmic ether, and has forced the question of its essence, its structure, and its motion into the foreground of monistic nature-philosophy. Only a few years ago the cosmic ether was to the majority of scientists an imponderable something, of which, strictly speaking, absolutely nothing was known, and which could be admitted provisionally only as a precarious working hypothesis. All this was changed when Heinrich Hertz (1888) demonstrated the nature of electrical energy, by his beautiful experiments establishing the conjecture of Faraday that light and heat, electricity and magnetism, are closely related phenomena of one single set of forces, and depend on transverse vibrations of the ether. Light itself—whatever else it be—is always and everywhere an electrical phenomenon. The ether itself is no longer hypothetical; its existence can at any moment be demonstrated by electrical and optical experiment. We know the length of the light wave and the electric wave. Indeed, some physicists believe that they can even determine approximately the density of ether. If by means of the air pump we remove from a bell-jar the atmospheric air (except an insignificant residue), the quantity of light within it remains unchanged; it is the vibrating ether we see.[8] These advances in our knowledge of the ether mean an immense gain for monistic philosophy. For they do away with the erroneous ideas of empty space and *actio in distans*; the whole of infinite space, in so far as it is not occupied by mass- atoms ("ponderable matter"), is filled by the ether. Our ideas of space and time are quite other than those taught by Kant a hundred years ago; the "critical" system of the great Königsberg philosopher exhibits in this respect, as well as in his teleological view of the organic world and in his metaphysics, dogmatic weaknesses of the most pronounced kind.[9] And religion itself, in its reasonable forms, can take over the ether theory as an article of faith, bringing into contradistinction the mobile cosmic ether as creating divinity, and the inert heavy mass as material of creation. From this successfully scaled height of monistic knowledge there open up before our joyously quickened spirit of research and discovery new and surprising prospects, which promise to bring us still nearer to the solution of the one great riddle of the world. What is the relation of this light mobile cosmic ether to the heavy inert "mass," to the ponderable matter which we chemically investigate, and which we can only think of as constituted of atoms? Our modern analytical chemistry remains for the present at a standstill, in presence of some seventy irreducible elements, or so-called primary substances. But the reciprocal relation of these elements, the affinity of their combinations, their spectroscopic behaviour, and so forth, make it in the highest degree probable that they are all merely historical products of an evolutionary process, having their origin in various dispositions and combinations of a varying number of original atoms.

To these original or mass-atoms—the ultimate discrete particles of inert "ponderable matter"—we can with more or less probability ascribe a number of eternal and inalienable fundamental attributes; they are probably everywhere in space, of like magnitude and constitution. Although possessing a definite finite magnitude, they are, by virtue of their very nature, indivisible. Their shape we may take to be spherical; they are inert (in the physical sense), unchangeable, inelastic, and impenetrable by the ether. Apart from the attribute of inertia, the most important characteristic of these ultimate atoms is their chemical affinity—their tendency to apply themselves to one another and combine into small groups in an orderly fashion. These fixed groups (fixed, that is to say, under the present physical conditions of existence of the earth) of primitive atoms are the atoms of the elements—the well-known "indivisible" atoms of chemistry. The qualitative, and, so far as our present empirical knowledge goes, unchangeable distinctions of our chemical elements are therefore solely conditioned by the varying number and disposition of the similar primitive atoms of which they are composed. Thus, for example, the atom of carbon (the real "maker" of the organic world) is in all probability a tetrahedron made up of four primitive atoms.

After Mendelejeff and Lothar Meyer had discovered (1869) the "periodic law" of the chemical elements, and founded on it a "natural system" of these elements, this important advance in theoretical chemistry was subsequently put to profitable use by Gustav Wendt from an evolutionary point of view. He endeavored to show that the various elements are products of evolution or of historically originating combinations of seven primary elements, and that these last again are historical products of one single primitive element This hypothetical original matter had been already designated by Crookes, in his *Genesis of the Elements*, as primary material or protyl.[10] The empirical proof of the existence of this original matter lying at the foundation of all ponderable material is perhaps only a question of time. Its discovery would probably realize the alchemists' hope of being able to produce gold and silver artificially out of other elements. But then arises the other great question: "How is this primary mass related to the cosmic ether? Do these two original substances stand in fundamental and eternal antithesis to one another? Or was it the mobile ether itself, perhaps, that originally engendered the heavy mass?"[11]

In answer to this great and fundamental question, various physical hypotheses have been put forward. But, like the various atomic theories of chemistry, they have not as yet been clearly established, and the same appears to me to be the case also with the ingenious hypothesis which the lecturer has unfolded to us with reference to the Influence of Space. As he himself rightly says, in all these endeavors after a philosophy of nature we are still, for the present, dealing with "scientific articles of faith," concerning the validity of which different persons, according to their subjective judgment and stage of culture, may have widely divergent views. I believe that the solution of these fundamental questions still lies as yet beyond the limits of our knowledge of nature, and that we shall be obliged, for a long time yet to come, to content ourselves with an "Ignoramus"—if not even with an "Ignorabimus."

The case is very different, however, if we turn from these atomistic element hypotheses and direct our attention to the historical conditions of the evolution of the world, as these have been revealed to us by the magnificent advances in our knowledge of nature which have been made within the last thirty years. An immense new territory has here been opened up to us in the realms of knowledge—a territory in which a series of most important problems, formerly held to be insoluble, has been answered in the most surprising manner.[12]

Among the triumphs of the human mind the modern doctrine of evolution takes a foremost place. Guessed at by Goethe a hundred years ago, but not expressed in definite form until formulated by Lamarck in the beginning of the present century, it was at last, thirty years ago, decisively established by Charles Darwin, his theory of selection filling up the gap which Lamarck in his doctrine of the reciprocal influence of heredity and adaptation had left open. We now definitely know that the organic world on our earth has been as continuously developed, "in accordance with eternal iron laws," as Lyell had in 1830 shown to be the case for the inorganic frame of the earth itself; we know that the innumerable varieties of animals and plants which during the course of millions of years have peopled our planet are all simply branches of one single genealogical tree; we know that the human race itself forms only one of the newest, highest, and most perfect offshoots from the race of the Vertebrates.

An unbroken series of natural events, following an orderly course of evolution according to fixed laws, now leads the reflecting human spirit through long æons from a primeval chaos to the present "order of the cosmos." At the outset there is nothing in infinite space but mobile elastic ether, and innumerable similar separate particles—the primitive atoms— scattered throughout it in the form of dust; perhaps these are themselves originally "points of condensation" of the vibrating "substance," the remainder of which constitutes the ether. The atoms of our elements arise from the grouping together in definite numbers of the primitive atoms or atoms of mass. As the Kant-Laplace nebular hypothesis has it, the rotating heavenly bodies separate themselves out from that vibrating primeval cloud. A single unit among many thousands of celestial bodies is our sun, with its planets, which originated by being centrifugally thrown off from it. Our insignificant earth is a single planet of our solar system; its entire individual life is a product of the sunlight. After the glowing sphere of the earth has cooled down to a certain degree, drops of fluid water precipitate themselves on the hardened crust of its surface—the first preliminary condition of organic life. Carbon atoms begin their organism-engendering activity, and unite with the other elements into plasma-combinations capable of growing. One small plasma-group oversteps the limits of cohesion and individual growth; it falls asunder into two similar halves. With this first moneron begins organic life and its most distinctive function, heredity. In the homogeneous plasma of the monera, a firmer central nucleus is separated from a softer outer mass; through this differentiation of nucleus and protoplasm arises the first organic cell. For a long time our planet was inhabited solely by such Protista or single-celled primitive creatures. From coenobia or social unions of these afterwards arose the lowest histones, multicellular plants and animals.

***B**y the sure help of the three great empirical "records of creation," paleontology, comparative anatomy, and ontogeny, the history of descent now leads us on step by step from the oldest Metazoa, the simplest pluricellular animals, up to man.[13] At the lowest root of the common genealogy of the Metazoa stand the Gastraeadae and Spongidæ; their whole body consists, in the simplest case, solely of a round digestive sac, the thin wall of which is formed by two layers of cells—the two primitive germinal layers. A corresponding germinal condition, the two-layered gastrula, occurs transitorily in the embryological history of all the other Metazoa, from the lowest Cnidaria and Vermes up to man. From the common stock of the Helminthes, or simple worms, there develop as independent main branches the four separate stems of the Molluscs, Star- fishes, Arthropods, and Vertebrates. It is only these last whose bodily structure and development in all essential respects coincide with those of man. A long series of lower aquatic Vertebrates (lancelets, lampreys, fishes) precedes the lung-breathing Amphibians, which appear for the first time in the Carboniferous period. The Amphibians are followed in the Permian period by the first Amniota, the oldest reptiles; from these develop later, in the Triassic period, the Birds on the one hand, and the Mammals on the other. That man in his whole bodily frame is a true mammal, becomes obvious as soon as the natural unity of this highest class of animals is recognized. The simplest comparison must have convinced the unprejudiced observer of the close constitutional relationship between man and the ape, which of all the Mammals comes nearest him. Comparative anatomy, with its deeper vision, showed that all differences in bodily structure between man and the Anthropoidæ (gorilla, chimpanzee, orangutan) are less important than the corresponding differences in bodily structure between these anthropoid apes and the lower apes. The phylogenetic significance of this fact, first emphasized by Huxley, is quite clear. The great question of the origin of the human race, or of "man's place in Nature," the "question of all questions," was then scientifically answered: "Man is descended from a series of ape-like Mammals." The descent of man (anthropogeny) discloses the long series of vertebrate ancestors, which preceded the late origin of this, its most highly developed offshoot.

The incalculable importance of the light cast over the whole field of human knowledge of nature by these results is patent to everyone. They are destined every year increasingly to manifest their transforming influence in all departments of knowledge, the more the conviction of their irrefragable truth forces its way. And it is only the ignorant or narrow-minded who can now doubt their truth. If, indeed, here and there, one of the older naturalists still disputes, the foundation on which they rest, or demands proofs which are wanting (as happened a few weeks ago on the part of a famous German pathologist at the Anthropological Congress in Moscow), he only shows by this that he has remained a stranger to the stupendous advances of recent biology, and above all of anthropogeny. The whole literature of modern biology, the whole of our present zoology and botany, morphology and physiology, anthropology and psychology, are pervaded and fertilized by the theory of descent.[14]

Just as the natural doctrine of development on a monistic basis has cleared up and elucidated the whole field of natural phenomena in their physical aspect, it has also modified that of the phenomena of mind, which is inseparably connected with the other. Our human body has been built up slowly and by degrees from a long series of vertebrate ancestors, and this is also true of our soul; as a function of our brain it has gradually been developed in reciprocal action and re-action with this its bodily organ. What we briefly designate as the "human soul," is only the sum of our feeling, willing, and thinking—the sum of those physiological functions whose elementary organs are constituted by the microscopic ganglion-cells of our brain. Comparative anatomy and ontogeny show us how the wonderful structure of this last, the organ of our human soul, has in the course of millions of years been gradually built up from the brains of higher and lower vertebrates. Comparative psychology teaches us how, hand in hand therewith, the soul itself, as function of the brain, has been developed. The last-named science teaches us also that a primitive form of soul-activity is already present even in the lowest animals, the single-celled primitive animals, Infusoria and Rhizopoda. Every scientific man who has long observed the life-activity of these single-celled Protista, is positively convinced that they also possess a soul; that this "cell-soul" also consists of a sum of sensations, perceptions, and volitions; the feeling, thinking, and willing of our human soul differ from these only in degree. In like manner there is present in the egg-cell (as potential energy) a hereditary cell-soul, out of which man, like every other animal, is developed.[15]

The first task of a truly scientific psychology will therefore be, not, as hitherto, idle speculation about an independent immaterial soul-existence and its puzzling temporary connection with the animal body, but rather the comparative investigation of the organs of the soul and the experimental examination of their psychical functions. For scientific psychology is a part of physiology, the doctrine of the functions and the life-activities of organisms. The psychology and psychiatry of the future, like the physiology and pathology of to-day, must take the form of a cellular study, and in the first instance investigate the soul-functions of the cells. Max Verworn, in his fine *Psycho-physiological Protistastudies*, has lately shown us what important disclosures such a cellular psychology can make, even in dealing with the lowest grades of organic life, in the single-celled Protista (especially Rhizopoda and Infusoria).

These same main divisions of soul-activity, which are to be met with in the single-celled organism,—the phenomena of irritability, sensation, and motion,—can be shown to exist in all multicellular organisms as functions of the cells of which their bodies are composed. In the lowest Metazoa, the invertebrate sponges and polyps, there are, just as in plants, no special soul- organs developed, and all the cells of the body participate more or less in the "soul-life." It is only in the higher animals that the soul-life is found to be localized and connected with special organs. As a consequence of division of labour, there have here been developed various sense-organs as organs of specific sensibility, muscles as organs of motion and volition, nerve-centers or ganglia as central co-ordinating and regulating organs. In the most highly developed families of the animal kingdom, these last come more and more into the foreground as independent soul-organs. In correspondence with the extraordinarily complicated structure of their central nervous system (the brain with its wonderful complex of ganglion- cells and nerve-fibers), the many-sided activity of such animals attains a wonderful degree of development.

It is only in these most highly-developed groups of the animal kingdom that we can with certainty establish the existence of those most perfect operations of the central nervous system, which we designate as consciousness. As we know, it is precisely this highest brain-function that still continues to be looked upon as a completely enigmatical phenomenon, and as the best proof for the immaterial existence of an immortal soul. It is usual at the same time to appeal to Du Bois-Reymond's well-known "Ignorabimus address on the Boundaries of Natural Knowledge" (1872). It was by a peculiar irony of fate that the famous lecturer of the Berlin Academy of Science, in this much-discussed address of twenty years ago, should be representing consciousness as an incomprehensible marvel, and as presenting an insuperable barrier to further advances of knowledge, at the very moment that David Friedrich Strauss, the greatest theologian of our century, was showing it to be the opposite. The clear-sighted author of *The Old Faith and the New* had already clearly perceived that the soul-activities of man, and therefore also his consciousness, as functions of the central nervous system, all spring from a common source, and, from a monistic point of view, come under the same category. The "exact" Berlin physiologist shut this knowledge out from his mind, and, with a short-sightedness almost inconceivable, placed this special neurological question alongside of the one great "world-riddle," the fundamental question of substance, the general question of the connection between matter and energy.[16]

As I long ago pointed out, these two great questions are not two separate "world-riddles." The neurological problem of consciousness is only a special case of the all comprehending cosmological problem, the question of substance. "If we understood the nature of matter and energy, we should also understand how the substance underlying them can under certain conditions feel, desire, and think." Consciousness, like feeling and willing, among the higher animals is a mechanical work of the ganglion-cells, and as such must be carried back to chemical and physical events in the plasma of these. And by the employment of the genetic and comparative method we reach the conviction that consciousness, and consequently reason also, is not a brain-function exclusively peculiar to man; it occurs also in many of the higher animals, not in Vertebrates only, but even in Articulates. Only in degree, through a higher stage of cultivation, does the consciousness of man differ from that of the more perfect lower animals, and the same is true of all other activities of the human soul.

By these and other results of comparative physiology our whole psychology is placed on a new and firm monistic basis. The older mystical conception of the soul, as we find it amongst primitive peoples, but also in the systems of the dualistic philosophers of to-day, is refuted by them. According to these systems, the soul of man (and of the higher animals) is a separate entity, which inhabits and rules the body only during its individual life, but leaves it at death. The widespread *Claviertheorie* (Piano Theory) compares the "immortal soul" to a pianist who executes an interesting piece —the individual life—on the instrument of the mortal body, but at death withdraws into the other world. This "immortal soul" is usually represented as an immaterial being; but in fact it is really thought of as quite material, only as a finer invisible being, aerial or gaseous, or as resembling the mobile, light, and thin substance of the ether, as conceived by modern physics. The same is true also for most of the conceptions which rude primitive peoples and the uneducated classes among the civilized races have, for thousands of years, cherished as to spectral "ghosts" and "gods." Serious reflection on the matter shows that here—as in modern spiritualism —it is not with really immaterial beings, but with gaseous, invisible bodies, that we are dealing. And further, we are utterly incapable of imagining a truly immaterial being. As Goethe clearly said, "matter can never exist or act apart from spirit, neither can spirit apart from matter."

As regards immortality, it is well known that this important idea is interpreted and applied in a great variety of ways. It is often made a reproach against our Monism that it altogether denies immortality; this, however, is erroneous. Rather do we hold it, in a strictly scientific sense, as an indispensable fundamental conception of our monistic philosophy of nature. Immortality in a scientific sense is conservation of substance, therefore the same as conservation of energy as defined by physics, or conservation of matter as defined by chemistry. The cosmos as a whole is immortal. It is just as inconceivable that any of the atoms of our brain or of the energies of our spirit should vanish out of the world, as that any other particle of matter or energy could do so. At our death there disappears only the individual form in which the nerve-substance was fashioned, and the personal "soul" which represented the work performed by this. The complicated chemical combinations of that nervous mass pass over into other combinations by decomposition, and the kinetic energy produced by them is transformed into other forms of motion.

On the other hand, the conception of a personal immortality cannot be maintained. If this idea is still widely held, the fact is to be explained by the physical law of inertia; for the property of persistence in a state of rest exercises its influence in the region of the ganglion-cells of the brain, as well as in all other natural bodies. Traditional ideas handed down through many generations are maintained with the greatest tenacity by the human brain, especially if, in early youth, they have been instilled into the childish understanding as indisputable dogmas. Such hereditary articles of faith take root all the more firmly, the further they are removed from a rational knowledge of nature, and enveloped in the mysterious mantle of mythological poesy. In the case of the dogma of personal immortality, there comes into play also the interest which man fancies himself to have in his individual future existence after death, and the vain hope that in a blessed world to come there is treasured up for him a compensation for the disappointed hopes and the many sorrows of his earthly life.

It is often asserted by the numerous advocates of personal immortality that this dogma is an innate one, common to all rational men, and that it is taught in all the more perfect forms of religion. But this is not correct. Neither Buddhism nor the religion of Moses originally contained the dogma of personal immortality, and just as little did the majority of educated people of classical antiquity believe it, at any rate during the highest period of Greek culture. The monistic philosophy of that time, which, five hundred years before our era, had reached speculative heights so remarkable, knew nothing of any such dogma. It was through Plato and Christ that it received its further elaboration, until, in the Middle Ages, it was so universally accepted, that only now and then did some bold thinker dare openly to gainsay it. The idea that a conviction of personal immortality has a specially ennobling influence on the moral nature of man, is not confirmed by the gruesome history of mediæval morals, and as little by the psychology of primitive peoples.[17]

*I*f any antiquated school of purely speculative psychology still continues to uphold this irrational dogma, the fact can only be regarded as a deplorable anachronism. Sixty years ago such a doctrine was excusable, for then nothing was accurately known either of the finer structure of the brain, or of the physiological functions of its separate parts; its elementary organs, the microscopic ganglion-cells, were almost unknown, as was also the cell- soul of the Protista; very imperfect ideas were held as to ontogenetic development, and as to phylogenetic there were none at all.

*T*his has all been completely changed in the course of the last half- century. Modern physiology has already to a great extent demonstrated the localization of the various activities of mind, and their connection with definite parts of the brain; psychiatry has shown that those psychical processes are disturbed or destroyed if these parts of the brain become diseased or degenerate. Histology has revealed to us the extremely complicated structure and arrangement of the ganglion-cells. But, for the settlement of this momentous question, the discoveries of the last ten years with regard to the more minute occurrences in the process of fertilization are of decisive importance. We now know that this process essentially consists simply in the copulation or fusion of two microscopical cells, the female egg-cell and the male sperm-cell. The fusion of the nuclei of these two sexual cells indicates with the utmost precision the exact moment at which the new human individual arises. The newly-formed parent-cell, or fertilized egg-cell, contains potentially, in their rudiments, all the bodily and mental characteristics which the child inherits from both parents. It is clearly against reason to assume an eternal and unending life for an individual phenomenon whose beginning in time we can determine to a hair's breadth, by direct observation. Judging of human spiritual life from a rational point of view, we can as little think of our individual soul as separated from our brain, as we can conceive the voluntary motion of our arm apart from the contraction of its muscles, or the circulation of our blood apart from the action of the heart.

Against this strictly physiological conception, as against our whole monistic view of the relations of energy and matter, of soul and substance, the reproach of "materialism" continues to be raised. I have repeatedly before now pointed out that this is an ambiguous party word which conveys absolutely nothing; its apparent opposite, "spiritualism," could quite easily be substituted for it. Every critical thinker, who is familiar with the history of philosophy, knows that, as systems change, such words assume the most varied meanings, In addition to this, the word "materialism" has the disadvantage of being liable to continual confusion between its theoretical and practical meanings, which two are totally distinct. Our conception of Monism, or the unity-philosophy, on the contrary, is clear and unambiguous; for it an immaterial living spirit is just as unthinkable as a dead, spiritless material; the two are inseparably combined in every atom. The opposed conception of dualism (or even pluralism in other anti- monistic systems) regards spirit and material, energy and matter, as two essentially different substances; but not a single empirical proof can be adduced to show that either of these can exist or become perceptible to us by itself alone.

In thus shortly indicating the far-reaching psychological consequences of the monistic doctrine of evolution, I trench at the same time upon a most important field, to which our lecturer in his address has more than once alluded—that of religion and the belief in God connected therewith. I am at one with him in the conviction that the formation of clear philosophical conceptions upon these fundamental matters of belief is of the highest importance, and I would therefore crave the permission of this assembly briefly to lay before it on this occasion a frank confession of faith. This monistic confession has the greater claim to an unprejudiced consideration, in that it is shared, I am firmly convinced, by at least nine-tenths of the men of science now living; indeed, I believe, by all men of science in whom the following four conditions are realized: (1) Sufficient acquaintance with the various departments of natural science, and in particular with the modern doctrine of evolution; (2) Sufficient acuteness and clearness of judgment to draw, by induction and deduction, the necessary logical consequences that flow from such empirical knowledge; (3) Sufficient moral courage to maintain the monistic knowledge, so gained, against the attacks of hostile dualistic and pluralistic systems; and (4) Sufficient strength of mind to free himself, by sound, independent reasoning, from dominant religious prejudices, and especially from those irrational dogmas which have been firmly lodged in our minds from earliest youth as indisputable revelations.

If from this unprejudiced point of view of the thinker, we compare the numerous religions of the various races of mankind, we shall be compelled, in the first instance, to put aside as untenable all those conceptions which stand in irreconcilable contradiction to those principles of our empirical knowledge of nature which are now clearly discerned and established by critical reasoning. We can thus at once set aside all mythological stories, all "miracles," and so-called "revelations," for which it is claimed that they have come to us in some supernatural way. All such mystical teachings are irrational, inasmuch as they are confirmed by no actual experience, but, on the contrary, are irreconcilable with the known facts which have been confirmed to us by a rational investigation of nature.

This is true alike of Christian and Mosaic, of Mohammedan and Indian legends. If now we thus lay aside the whole mass of mystical dogmas and transcendental revelations, there is left behind, as the precious and priceless kernel of true religion, the purified ethic that rests on rational anthropology.[18]

Among the numerous and varied forms of religion which, in the course of the past ten thousand years and more, have been evolved from the crudest prehistoric beginnings, the foremost rank undoubtedly belongs to those two forms which still continue to be the most widely accepted among civilized races—the older Buddhism and the younger Christianity. The two have very many features in common, alike in their mythology and in their ethics; indeed, a considerable part of Christianity has come directly from Indian Buddhism, just as another part is drawn from the Mosaic and Platonic systems. But, looked at from the point of view of our present stage of culture, the ethic of Christianity appears to us much more perfect and pure than that of any other religion. We must, it is true, hasten to add that it is exactly the weightiest and noblest principles of Christian ethic—brotherly love, fidelity to duty, love of truth, obedience to law—that are by no means peculiar to the Christian faith as such, but are of much older origin. Comparative psychology proves that these ethical principles were more or less recognized and practiced by much older civilized races thousands of years before Christ.

Love remains the supreme moral law of rational religion, the love, that is to say, that holds the balance between egoism and altruism, between self- love and love of others. "Do to others as you would they should do to you." This natural and highest command had been taught and followed thousands of years before Christ said: "Thou shalt love thy neighbor as thyself." In the human family this maxim has always been accepted as self-evident; as ethical instinct it was an inheritance derived from our animal ancestors. It had already found a place among the herds of Apes and other social Mammals; in a similar manner, but with a wider scope, it was already present in the most primitive communities and among the hordes of the least advanced savages. Brotherly love—mutual support, succor, protection, and the like—-had already made its appearance among gregarious animals as a social duty; for without it the continued existence of such societies is impossible. Although at a later period, in the case of man, these moral foundations of society came to be much more highly developed, their oldest prehistoric source, as Darwin has shown, is to be sought in the social instincts of animals. Among the higher Vertebrates (dogs, horses, elephants, etc.), as among the higher Articulates (ants, bees, termites, etc.) also, the development of social relations and duties is the indispensable condition of their living together in orderly societies. Such societies have for man also been the most important instrument of intellectual and moral progress.

Beyond all doubt the present degree of human culture owes in great part its perfection to the propagation of the Christian system of morals and its ennobling influence, although the great value of this has been impaired, often in the most deplorable manner, by its association with untenable myths and so-called "revelations." How little these last contribute to the perfection of the first, can be seen from the acknowledged historical fact that it is just orthodoxy and the hierarchical system based on it (especially that of the Papacy) that has least of all striven to fulfill the precepts of Christian morality; the more loudly they preach it in theory, the less do they themselves fulfill its commands in practice.

It is, moreover, to be borne in mind that another and very considerable portion of our modern culture and morality has been developed quite independently of Christianity, mainly through continual study of the highly-elaborated mental treasures of classical antiquity. The thorough study of Greek and Roman classics has at least contributed much more to it than that of the Christian Church fathers. To this we must now add, in our own century (rightly called the "century of the natural sciences"), the immense advance in the higher culture which we owe to a purified knowledge of nature and to the monistic philosophy founded upon this. That these must also exercise an advancing and ennobling influence cannot be doubted, and has already been shown by many eminent authors (Spencer, Carneri, and others) in the course of the last thirty years.

Against this monistic ethic founded on a rational knowledge of nature, it has been objected that it is fitted to undermine existing civilization, and especially that it encourages the subversive aims of social democracy. This reproach is wholly unjustified. The application of philosophical principles to the practical conditions of life, and in particular to social and political questions, can be made in the most various ways. Political "free-thinking," so called, has nothing whatever to do with the "freedom of thought" of our monistic natural religion. Moreover, I am convinced that the rational morality of monistic religion is in no way contrary to the good and truly valuable elements of the Christian ethic, but is destined in conjunction with these to promote the true progress of humanity in the future.

With Christian mythology and the special form of theistic belief associated with it the case is different. In so far as that belief involves the notion of a "personal God," it has been rendered quite untenable by the recent advances of monistic science. But, more than this, it was shown more than two thousand years ago, by eminent exponents of the monistic philosophy, that the conception of a personal God, creator and ruler of the world, does not give the slightest help toward a truly rational view of the world. For even if the question of "creation," in the ordinary and trivial sense of the term, be answered by referring it to the miraculous agency of a creator working according to plan apart from the world, there immediately arises upon that the new inquiry: "Whence comes this personal God? What was He doing before creation? And whence did He derive the material for it?" and such like questions. The antiquated conception of an anthropomorphic personal God is destined, before the present century is ended, to drop out of currency throughout the entire domain of truly scientific philosophy; the corresponding conception of a personal devil— even as late as last century connected with the former and very generally accepted—has already been given up once for all by all persons of education.

Let it be noted, however, in passing, that the amphitheism which believes in God and devil alike is much more compatible with a rational explanation of the world than pure monotheism. The purest form of this is perhaps the amphitheism of the Zend religion of Persia, which Zoroaster (or Zarathustra, the "Golden Star") founded two thousand years before Christ. Here Ormuzd, the god of light and goodness, stands everywhere in conflict with Ahriman, the god of darkness and evil. The continual conflict between a good and an evil principle was personified in a similar manner in the mythology of many other amphitheistic religions: in the old Egyptian, the good Osiris was at war with the evil Typhon; in the old Indian, Vishnu the sustainer with Shiva the destroyer, and so forth.

If we really must retain the conception of a personal God as the key to our view of the universe, then this amphitheism can explain the sorrows and defects of this world very simply, as being the work of the evil principle or devil. Pure monotheism, on the contrary, as represented in the religions of Moses and Mohammed in their original form, has no rational explanation of these to offer. If their "one God" is really the absolutely good, perfect being they proclaim, then the world which he has created must also be perfect. An organic world so imperfect and full of sorrows as exists on this earth he could not possibly have contrived.

These considerations gain in force when we advance to the deeper knowledge of nature acquired by modern biology; here it was Darwin, especially, who thirty-three years ago opened our eyes by his doctrine of the struggle for existence, and his theory of selection founded upon it. We now know that the whole of organic nature on our planet exists only by a relentless war of all against all. Thousands of animals and plants must daily perish in every part of the earth, in order that a few chosen individuals may continue to subsist and to enjoy life. But even the existence of these favoured few is a continual conflict with threatening dangers of every kind. Thousands of hopeful germs perish uselessly every minute. The raging war of interests in human society is only a feeble picture of the unceasing and terrible war of existence which reigns throughout the whole of the living world. The beautiful dream of God's goodness and wisdom in nature, to which as children we listened so devoutly fifty years ago, no longer finds credit now—at least among educated people who think. It has disappeared before our deeper acquaintance with the mutual relations of organisms, the advancement of oecology and sociology, and our knowledge of parasite life and pathology.

All these sad but insuperable facts—truly the dark side of nature—are made intelligible to religious faith by amphitheism; they are the "works of the devil," who opposes and disturbs the perfect moral order in the world of the "good God." For pure monotheism which knows only one God, one perfect highest being, they remain unintelligible. If, with a monotheistic creed, any one still continues to talk of the moral order of the world, he in so doing shuts his eyes to the undeniable facts of history, both natural and civil.

*I*n view of these considerations, it is hard to understand how the large majority of the so-called educated classes can persevere, on the one hand, in declaring belief in a personal God to be an indispensable principle of religion, and, on the other hand, in at the same time rejecting the belief in a personal devil as an exploded superstition of the Middle Ages. This inconsistency on the part of educated Christians is all the more incomprehensible and censurable, inasmuch as both dogmas in equal degree form an integral part of the Christian creed. The personal devil, as "Satan," "the Tempter," "the Destroyer," and so forth, undeniably plays a most important part in the New Testament, though not met with in the earlier portions of the Old. Our great reformer, Martin Luther himself, who "sent to the devil" so many antiquated dogmas, was unable to rid himself of the conviction of the real existence and personal enmity of Beelzebub; we have only to think of the historical ink-spot at Wartburg! Moreover, our Christian art, in many thousands of paintings and other representations, has exhibited Satan in corporeal form just as realistically as it has the three "Divine Persons," about whose "hypostatical union" human reason has for eighteen hundred years been tormenting itself in vain. The deep impression made by such concrete representations, a million times repeated, especially on childish understandings, is usually under-estimated as to its tremendous influence; to it certainly is in large measure to be attributed the fact that irrational myths of such a kind, under the mask of "doctrines of faith," continue to hold their ground in spite of all protests of reason.

Liberal-minded Christian theologians have, it is true, often sought to eliminate the personal devil from Christian teaching, representing him as merely the personification of falsehood, the spirit of evil. But with equal right we must in that case substitute for a personal God the personified idea of truth, the Spirit of Goodness. To such a representation no objection can be made; rather do we recognize in it a bridge connecting the dim wonderland of religious poesy with the luminous realms of clear scientific knowledge.

The monistic idea of God, which alone is compatible with our present knowledge of nature, recognizes the divine spirit in all things. It can never recognize in God a "personal being," or, in other words, an individual of limited extension in space, or even of human form. God is everywhere. As Giordano Bruno has it: "There is one spirit in all things, and nobody is so small that it does not contain a part of the divine substance whereby it is animated." Every atom is thus animated, and so is the ether; we might, therefore, represent God as the infinite sum of all natural forces, the sum of all atomic forces and all ether-vibrations. It comes virtually to the same thing when (as was done here by a speaker on a former occasion) God is defined as "the supreme law of the universe," and the latter is represented as the "working of universal space." In this most important article of belief it matters not as to the name but as to the unity of the underlying idea; the unity of God and the world; of spirit and nature. On the other hand, "homotheism," the anthropomorphic representation of God, degrades this loftiest cosmic idea to that of a "gaseous vertebrate."[19]

Of the various systems of pantheism which for long have given expression more or less clearly to the monistic conception of God, the most perfect is certainly that of Spinoza. To this system, as is well known, Goethe also paid the tribute of his highest admiration and approval. Of other, eminent men who have given a similar pantheistic form to their natural religion, we shall here mention only two of the greatest poets and students of man, Shakespeare and Lessing; two of the greatest German rulers, Frederick II. of Hohenstaufen and Frederick II. of Hohenzollern; two of the greatest scientists, Laplace and Darwin. In adding our own pantheistic confession to that of these great and untrammeled spirits, let it only be noted further, that it has received an empirical confirmation, never before imagined, through the wonderful advances of natural knowledge within the last thirty years.

The charge of atheism which still continues to be leveled against our pantheism, and against the monism which lies at its root, no longer finds a response among the really educated classes of the present day. It is true that not so very long ago the German Imperial Chancellor, in the Prussian Chamber of Deputies, found it in him to put forward such an alternative as this: "Either the Christian or the atheistic view of the world"; this in the defense of a most objectionable law, designed to hand over our school training, tied hand and foot, to the papal hierarchy. The vast distance which separates the last-named degenerate outgrowth of the Christian religion from pure primitive Christianity is not greater than that which separates those mediæval alternatives from the cultured religious consciousness of the present day. To one who regards as true exercises of Christian religion the adoration of old clothes and wax dolls, or the thoughtless repetition of masses or rosaries, who believes in wonder-working relics, and purchases pardon for his sins by means of indulgence-money or Peter's pence, we willingly concede the claim to possess the "only saving religion"; but with such fetish-worshippers we will willingly submit to be ranked as "atheists."

*I*n like case with the charge of atheism and irreligion are those so often heard against monism, that it destroys the poetry of life and fails to satisfy the spiritual wants of human nature; we are told, in particular, that aesthetics—certainly a most important department both in theoretical philosophy and in practical life—is prejudiced by a monistic philosophy. But David Friedrich Strauss, one of our subtlest exponents of aesthetics and also one of our noblest writers, has already refuted such a charge; and shown how, on the contrary, the care for poetry and the cultivation of the beautiful are in the "new faith" called upon to play a still greater part than ever. My present hearers, at once investigators and lovers of nature, do not need to be told that every new insight which we obtain into the secrets of nature at the same time also kindles our souls, affords new material for imagination to work on, and enlarges our perception of the beautiful. To convince ourselves how closely all these noblest spiritual activities of man hang together, how intimately the knowledge of truth is bound up with the love of goodness and veneration of the beautiful, it will be enough to mention a single name, Germany's greatest genius—Wolfgang Goethe.

*I*f the perception of the aesthetic significance of our monistic nature- religion, as well as of its ethical value, has hitherto so little pervaded the educated classes, this is due chiefly to the defects of our school training. It is true that in the course of the last few decades an infinite deal has been spoken and written about school reform and the principles of education; but of any real progress there is as yet but little trace. Here also reigns the physical law of inertia; here also—and more especially in German schools —the scholasticism of the Middle Ages exhibits a power of inertia, against which any rational reform of education must laboriously contest every inch of ground. In this important department also, a department on which hangs the weal or woe of future generations, matters will not improve till the monistic doctrine of nature is accepted as the essential and sure foundation.

The school of the twentieth century, flourishing anew on this firm ground, shall have to unfold to the rising youth not only the wonderful truths of the evolution of the cosmos, but also the inexhaustible treasures of beauty lying everywhere hidden therein. Whether we marvel at the majesty of the lofty mountains or the magic world of the sea, whether with the telescope we explore the infinitely great wonders of the starry heaven, or with the microscope the yet more surprising wonders of a life infinitely small, everywhere does Divine Nature open up to us an inexhaustible fountain of aesthetic enjoyment. Blind and insensible have the great majority of mankind hitherto wandered through this glorious wonderland of a world; a sickly and unnatural theology has made it repulsive as a "vale of tears." But now, at last, it is given to the mightily advancing human mind to have its eyes opened; it is given to it to show that a true knowledge of nature affords full satisfaction and inexhaustible nourishment not only for its searching understanding, but also for its yearning spirit.

Monistic investigation of nature as knowledge of the true, monistic ethic as training for the good, monistic aesthetic as pursuit of the beautiful— these are the three great departments of our monism: by the harmonious and consistent cultivation of these we effect at last the truly beatific union of religion and science, so painfully longed after by so many to-day. The True, the Beautiful, and the Good, these are the three august Divine Ones before which we bow the knee in adoration; in the unforced combination and mutual supplementing of these we gain the pure idea of God.[20] To this "triune" Divine Ideal shall the coming twentieth century build its altars.

Ten years ago I was present at the celebration of the third centenary of the university of Würzburg, which forty years ago I had entered as a medical student. The festal address on that occasion was delivered in the university church by the then rector, the distinguished chemist, Johannes Wislicenus. His concluding words were: "God, the Spirit of Goodness and of Truth, grant it." I now add, "and the Spirit of Beauty." It is in this sense that I also, on this commemorative occasion, dedicate to you my best wishes. May the investigation of nature's secrets flourish and prosper in this corner of our Thüringian land also, and may the fruits of knowledge, ripening here in Altenburg, contribute no less to the culture of the spirit and to the advancement of true religion, than those which three hundred and seventy years ago the great reformer, Martin Luther, brought to the light of day in another corner of Thüringen, on the Wartburg at Eisenach.

Between Wartburg and Altenburg, on the northern border of Thüringen, lies Weimar, the classical City of the Muses, and, close by it, our national university of Jena. I regard it as a good omen that precisely at this moment a rare celebration should have called together in Weimar the most illustrious patrons of the university of Jena, the defenders of free research and free teaching.[21] In the hope that the defense and promotion of these may still be continued, I conclude my monistic Confession of Faith with the words: "May God, the Spirit of the Good, the Beautiful, and the True, be with us."

Notes

1:

[*Scientific Articles of Faith*. In Professor Schlesinger's address (delivered on 9th October at Altenburg) on this subject he rightly called attention to the limits of knowledge of nature (in Kant's sense of the terms) imposed upon us by the imperfection of our perceptive organs. The gaps which the empirical investigation of nature must thus leave in science, can, however, be filled up by hypotheses, by conjectures of more or less probability. These we cannot indeed for the time establish on a secure basis; and yet we may make use of them in the way of explaining phenomena, in so far as they are not inconsistent with a rational knowledge of nature. Such rational hypotheses are scientific articles of faith, and therefore very different from ecclesiastical articles of faith or religious dogmas, which are either pure fictions (resting on no empirical evidence), or simply irrational (contradicting the law of causality). As instances of rational hypotheses of first-rate importance may be mentioned our belief in the oneness of matter (the building up of the elements from primary atoms), our belief in equivocal generation, our belief in the essential unity of all natural phenomena, as maintained by monism (on which compare my *General Morphology, vol*. i. pp. 105, 164, etc., also my *Natural History of Creation*, 8th ed., 1889, pp. 21, 360, 795). As the simpler occurrences of inorganic nature and the more complicated phenomena of organic life are alike reducible to the same natural forces, and as, further, these in their turn have their common foundation in a simple primal principle pervading infinite space, we can regard this last (the cosmic ether) as all-comprehending divinity, and upon this found the thesis: "Belief in God is reconcilable with science." In this pantheistic view, and also in his criticism of a one-sided materialism, I entirely agree with Professor Schlesinger, though unable to concur with him in some of his biological, and especially of his anthropological, conclusions (*cf.* his article on "Facts and Deductions derived from the Action of Universal Space" *Mittheilungen aus dem Osterlande*, Bd. v., Altenburg, 1892).]

2:

[*Unity of Nature*. I consider the fundamental unity of inorganic and organic nature, as well as their genetic relation, to be an essential axiom of monism. I particularly emphasize this "article of faith" here, as there are still scientists of repute who contest it. Not only is the old mystical "vital power" brought back upon the stage again from time to time, but even the "miraculous" origin of organic life out of "dead" inorganic nature is often brought up still against the doctrines of evolution, as an insoluble riddle—as one of Du Bois-Reymond's "seven riddles of the world" (see his *Discourse on Leibnitz*, 1880). The solution of this "transcendent" riddle of the world, and of the allied question of archigony (equivocal generation, in a strictly defined meaning of the term), can only be reached by a critical analysis and unprejudiced comparison of matter, form, and energy in inorganic and organic nature. This I have already done (1866) in the second book of my *General Morphology* (vol. i. pp. 109-238): "General Researches as to the Nature and First Beginning of Organisms, their Relation to things Inorganic, and their Division into Plants and Animals."] A short résumé of this is contained in Lecture XV. of my *Natural History of Creation* (8th ed., pp. 340-370). The most serious difficulties which formerly beset the monistic view there given may now be held to have been taken out of the way by recent discoveries concerning the nature of protoplasm, the discovery of the Monera, the more accurate study of the closely-related single-celled Protista, their comparison with the ancestral cell (or fertilized egg-cell), and also by the chemical carbon-theory. (See my "Studies on Monera and other Protista," in the *Jenaische Zeitschrift für Naturwissenschaft*, vols. iv. and v., 1868-1870; also Carl Naegeli, *Mechanisch-physiologische Begründung der Abstammungslehre*, 1884.)]

3:

[*Religion in the Lower Animals*. We cannot fail to recognize in the more highly developed of our domestic animals (especially in dogs, horses, and elephants) some first beginnings of those higher brain-functions which we designate as reason and consciousness, religion and morality; they differ only in degree, not in kind, from the corresponding mental activities of the lowest human races. If, like the dogs, the apes, and especially the anthropoids, had been for thousands of years domesticated and brought up in close relation with civilized man, the similarity of their mental activities to those of man would undoubtedly have been much more striking than it is. The apparently deep gulf which separates man from these most highly-developed mammals "is mainly founded on the fact that in man several conspicuous attributes are united, which in the other animals occur only separately, viz. (1) The higher degree of differentiation of the larynx (speech), (2) brain (mind), and (3) extremities; and (4) the upright posture. It is merely the happy combination of these important animal organs and functions at a higher stage of evolution that raises the majority of mankind so far above all lower animals" (*General Morphology*, 1866, vol. ii. p. 430).]

4:

[*Inheritance of Acquired Characters*. As the controversy on this important question is still unsettled, special attention may here be called to the valuable data for arriving at a decision which are afforded precisely by the development of instincts among the higher animals, and of speech and reason in man. "The inheritance of characters acquired during the life of the individual, is an indispensable axiom of the monistic doctrine of evolution." "Those who, with Weismann and Galton, deny this, entirely exclude thereby the possibility of any formative influence of the outer world upon organic form" (*Anthropogenie*, 4th ed., pp. xxiii., 836; see, further, the works there referred to of Eimer, Weismann, Ray-Lankester, etc.; also Ludwig Wilser's *Die Vererbung der geistigen Eigenschaften*, Heidelberg, 1892).]

5:

[*Theosophical System of Nature*. Of all the modern attempts of dualistic philosophy to establish the knowledge of nature on a theological basis (that of Christian monotheism), the *Essay on Classification* of Louis Agassiz is by far the most important,—in strictness, indeed, is the only one worthy of mention. (On this see my *Natural History of Creation*, Lect. III., also "Aims and Methods of the Modern Embryology," 1875, *Jena Zeitschr. für Naturw., Bd. x., Supplement.*)]

6:

[*Darwin and Copernicus*. This is the title of an address delivered by Du Bois-Reymond on 25th January 1883, in the Berlin Academy of Sciences, and afterwards published in his *Collected Addresses* (*vol.* ii. 1887). As the author himself mentions in a note (p. 500) that this gave rise, "most unmeritedly," to great excitement, and called down upon him the violent attacks of the clerical press, I may be allowed to point out here that it contained nothing new, I myself, fifteen years previously, in my lectures on "The Origin and Genealogy of the Human Race," having carried out in detail the comparison between Darwin and Copernicus, and the service rendered by these two heroes in putting an end to the anthropocentric and geocentric views of the world. (See the Third Series in Virchow and Holtzendorff's *Collection of Popular Scientific Lectures*, Nos. 53 and 54, 1868, 4th ed., 1881.) When Du Bois-Reymond says, "For me, Darwin is the Copernicus of the organic world," I am the more pleased to find that he agrees (partly in identical words) with my way of thinking, as he himself, quite unnecessarily, takes up an attitude of opposition towards me. The same is the case with regard to the explanation of innate ideas by Darwinism, which he has attempted in his address (1870) on "Leibnitzian Ideas in Modern Science" (vol. i. of the *Collected Addresses*). Here also he is most agreeably at one with me in what, four years before, I had elaborated in my *General Morphology* (vol. ii. p. 446), and in my *Natural*

History of Creation (1868). "The laws of heredity and adaptation explain to us how it is that *à priori* ideas have been developed out of what was originally *à posteriori* knowledge," etc. I cannot fail to be highly flattered in being able in these last days to greet the renowned orator of the Berlin Academy as a friend and patron of the *Natural History of Creation*, which he had previously designated a bad romance. But his winged words are not on that account to be forgotten, that "the genealogical trees of phylogeny are about as much worth as, in the eyes of the historical critic, are those of the Homeric heroes" (*Darwin versus Galiani*, 1876).]

7:

[*The Law of the Conservation of Substance*. Strictly taken, this belongs also to "scientific articles of faith," and could stand as the first article of our "monistic religion." Physicists of the present day, it is true, generally (and correctly) regard their "law of the conservation of energy" as the immovable foundation of all their science (Robert Mayer, Helmholtz), just as in like manner chemists so regard their fundamental law of the "conservation of matter" (Lavoisier). Skeptical philosophers could, however, raise certain objections to either of these fundamental laws with as much success as against their combination into the single superior law of the "conservation of substance." As a matter of fact, dualistic philosophy still attempts to raise such objections, often under the guise of cautious criticism. The skeptical (in part also purely dogmatic) objections have a semblance of justification only in so far as they relate to the fundamental problem of substance, the primary question as to the connection between matter and energy. While freely recognizing the presence of this real "boundary of natural knowledge," we can yet, within this boundary, apply quite universally the "mechanical law of causality." The complicated "phenomena of mind," as they are called (more especially consciousness), fall under the "law of the conservation of substance" just as strictly as do the simpler mechanical processes of nature dealt with in inorganic physics and chemistry. Compare note 16.]

8:

[*The Ether.* In a thoughtful lecture on the relations between light and electricity at the sixty-second Congress of German naturalists and physicians in Heidelberg in 1889, Heinrich Hertz explains the scope of his brilliant discovery: "Thus the domain of electricity extends over the whole of nature. It comes nearer to ourselves; we learn that we actually possess an electric organ, the eye. Here we are brought face to face with the question as to unmediated *actio in distans*. Is there such a thing? Not far off from this, in another direction, lies the question of the nature of electricity. And immediately connected therewith arises the momentous and primary question as to the nature of the ether, of the properties of the medium that fills all space, its structure, its rest or motion, its infinitude or finitude. It becomes every day more manifest that this question rises above all others, that a knowledge of what the ether is would reveal to us not only the nature of the old 'imponderables,' but also of the old 'matter' itself and its most essential properties, weight and inertia. Modern physics is not far from the question whether everything that exists is not created from the ether." This question is already being answered in the affirmative by some monistic physicists, as, for example, by J. G. Vogt in his most suggestive work on *The Nature of Electricity and Magnetism*, on *The Basis of the Conception of a Single Substance* (Leipsic, 1891). He regards the atoms of mass (the primal atoms of the kinetic theory of matter) as individualized centres of concentration of the continuous substance that uninterruptedly fills all space; the mobile elastic part of this substance between the atoms, and universally distributed, is—the ether. Georg Helm in Dresden, on the basis of mathematico-physical experiments, had already at an earlier date arrived at the same conclusions; in his treatise on "Influences at a Distance mediated by the Ether" (*Annalen der Physik und Chemie*, 1881, Bd. xiv.), he shows that it requires only the postulate of one particular kind of matter, the ether, to explain influence at a distance and radiation; that is, as regards these phenomena, all the qualities ascribable to matter, except that of motion, are of no account; in other words, that in thinking of the ether we simply require to think of it as "the mobile."]

9:

[*Kant and Monism*. As recent German philosophy has in a large measure returned to Kant, and in some cases even deified as "infallible" the great Königsberg philosopher, it may be well here to point out once more that his system of critical philosophy is a mixture of monistic and dualistic ingredients. His critical principles of the theory of knowledge will always remain of fundamental importance: his proof that we are unable to know the essential and profoundest essence of substance, the "thing in itself" (or "the combination of matter and energy"); that our knowledge remains subjective in its nature; that it is conditioned by the organization of our brain and sensory organs, and can therefore only deal with the phenomena which our experience of the outer world affords us. But within these "limits of human knowledge" a positive monistic knowledge of nature is still possible, in contrast to all dualistic and metaphysical fantasies. One such great fact of monistic knowledge was the mechanical cosmogony of Kant and Laplace, the "Essay on the Constitution and Mechanical Origin of the Universe, according to the Principles of Newton" (1755). In the whole field of our knowledge of inorganic nature, Kant held firmly to the monistic point of view, allowing mechanism alone as the real explanation of the phenomena. In the science of organic nature also, on the other hand, he held monism to be valid indeed, yet insufficient; here he considered it necessary to call in the aid of final as well as of efficient causes. (*Cf*. the fifth lecture of my *Natural History of Creation* on "The Evolution-Theory of Kant and Lamarck"; also Albrecht Rau's *Kant und die Naturforschung: Eine Prüfung der Resultate des idealistischen Kritikismus durch den realistischen Kosmos*, vol. ii., 1886.) Once thus on the downgrade of dualistic teleology, Kant afterwards arrived at his untenable metaphysical views of "God, Freedom, and Immortality." It is probable that Kant would have escaped these errors if he had had a thorough anatomical and physiological training. The natural sciences were, indeed, at that time truly in their infancy. I am firmly convinced that Kant's system of critical philosophy would have turned out quite otherwise from what it was, and purely monistic, if he had had at his disposal the then unsuspected treasures of empirical natural knowledge which we now possess.]

10:

[*Atoms and Elements*. The evidences, numerous and important, for the composite nature of our empirical elements, have lately been compendiously discussed by Gustav Wendt in his treatise, *Die Entwicklung der Elemente: Entwurf zu einer biologischen Grundlage fur Chemie und Physik*[I] (Berlin, 1891); compare also Wilhelm Freyer's *Die organischen Elemente und ihre Stellung im System*[II] (Wiesbaden, 1891), Victor Meyer's *Chemische Probleme der Gegenwart*[III] (Heidelberg, 1890), and W. Crookes's *Genesis of the Elements*. For the different views as to the nature of the atom, see Philip Spiller on "The Doctrines of Atoms" in *Die Urkraft des Weltalls nach ihrem Wesen und Wirken auf allen Naturgebieten[IV]* (Berlin, 1886), (1. The philosophy of nature; 2. The doctrine of the ether; 3. The ethical side of the science of nature). For the constitution of the elements out of atoms, see A. Turner, Die Kraft und Masse im Raume[V] (Leipsic, 3rd ed., 1886), (1. On the nature of matter and its relationships; 2. Atomic combinations; 3. The nature of the molecules and their combinations. Theory of crystallization).

Note I "The Development of the Elements: an Essay towards a Biological Basis for Chemistry and Physics."
Note II "The Organic Elements and their Place in the System."
Note III "Chemical Problems of the Day."
Note IV "The Primary Force of the Universe, its Nature and Action."
Note V "Force and Matter in Space."]

11:

[*World-Substance*. The relation of the two fundamental constituents of the cosmos, ether and mass, may perhaps be made apparent, in accordance with one out of many hypotheses, by the following, partly provisional, scheme.]

World (=Substance=Cosmos)
(*Nature as knowable by Man.*)

Æther (="spirit") (mobile or active substance).
Property of Vibration.

Chief Functions: Electricity, Magnetism, Light, Heat.

Structure: dynamical; continuous, elastic substance, not composed of atoms (*?*)

Theosophical: "God the Creator" (always in motion).

"Influence of space."

Mass (="body") (inert or passive substance).
Property of Inertia.

Chief Functions: Gravity, Inertia, Chemical Affinity.

Structure: atomic, discontinuous, inelastic substance,
composed of atoms (*?*)

Theosophical: "Created world" (passively formed).

"Products of space condensation."

12:

[*General doctrine of Evolution*. The fundamental importance of the modern doctrine of evolution, and of the monistic philosophy based upon it, is clearly evidenced by the steady increase of its copious literature. I have cited the most important treatises on this subject in the new (eighth) edition of my *Natural History of Creation* (1889). Compare, specially, Carus Sterne (Ernst Krause), *Werden und Vergehen: Eine Entwicklungsgeschichte des Naturganzen in gemeinverständlicher Fassung*[VI] (3rd ed., Berlin, 1886); Hugo Spitzer, *Beiträge zur Descendenztheorie und zur Methodologie der Naturwissenschaft* (Graz, 1886);[VII] Albrecht Ran, *Ludwig Feuerbach's Philosophie der Naturforschung und die philosophische Kritik der Gegenwart* (Leipsic, 1882); [VIII] Hermann Wolff, *Kosmos: Die Weltentwicklung nach monitisch-psychologischen Principien auf Grundlage der exacten Naturforschung* (Leipsic, 1890).[IX]

Note VI "Growth and Decay: a Popular History of the Development of the Cosmos."

Note VII "Contributions towards a Theory of Descent, and towards a Methodology of the Sciences of Nature."

Note VIII "Ludwig Feuerbach's Philosophy of Science, and the Philosophical Criticism of the Present Time."

Note IX "Cosmos: The Development of the Cosmos according to Monistic Principles on the Basis of Exact Science."]

13:

[*History of Descent*. The idea and the task of phylogeny, or the history of descent, I first defined in 1866, in the sixth book of my *General Morphology* (*vol.* ii. pp. 301-422), and the substance of this, as well as an account of its relation to ontogeny or history of development, is set forth in a popular form in Part II. of my *Natural History of Creation* (8th ed., Berlin, 1889). A special application of both these divisions of the history of evolution to man, is attempted in my *Anthropogenie* (4th ed.), revised and enlarged, 1891: Part I. History of development. Part II. History of descent.]

14:

[*Opponents of the Doctrine of Descent*. Since the death of Louis Agassiz (1873), Rudolf Virchow is regarded as the sole noteworthy opponent of Darwinism and the theory of descent; he never misses an opportunity (as recently in Moscow) of opposing it as "unproved hypothesis." See as to this my pamphlet, *Freedom in Science and in Teaching*, a reply to Virchow's address at Munich on "Freedom of Science in the Modern State" (Stuttgart, 1878; Eng. tr., 1892).]

15:

[*Cellular Psychology*. See on this my paper on "Cell-souls and Soul-cells," in the *Deutsche Rundschau* (July 1878), reprinted in Part 1, of *Collected Popular Lectures*; also "The Cell-soul and Cellular Psychology" in my discourse on *Freedom in Science and Teaching* (Stuttgart, 1878; Eng. tr., 1892, p. 46); *Natural History of Creation* (8th ed., pp. 444, 777); and *Descent of Man* (4th ed., pp. 128, 147). See also, Max Verworn, *Psycho-physiologische Protisten-Studien* (Jena, 1889), and Paul Carus, *The Soul of Man: An Investigation of the Facts of Physiological and Experimental Psychology* (Chicago, 1891). Among recent attempts to reform psychology on the basis of evolutionary doctrine in a monistic sense, special mention must be made of Georg Heinrich Schneider's *Der thierische Wille: Systematische Darstellung und Erklärung der thierischen Triebe und deren Entstehung, Entwickelung und Verbreitung im Thierreiche als Grundlage zu einer vergleichenden Willenslehre*[X] (Leipsic, 1880). Compare also his supplementary work, entitled *Der menschliche Wille vom Standpunkte der neuen Entwickelungstheorie*[XI] (1882); also the *Psychology of Herbert Spencer* and the new edition of Wilhelm Wundt's *Menschen-und Thierseele[XII]* (Leipsic, 1892).

Note X "Will in the Lower Animals: a Systematic Exposition and Explanation of Animal Instincts, and their Origin, Development, and Difference in the Animal Kingdom, as Basis of a Comparative Doctrine of Volition."

Note XI "The Human Will from the Standpoint of the Modern Theory of Evolution."

Note XII "Soul in Man and Brute."

16:

[*Consciousness.* The antiquated view of Du Bois-Reymond (1872)—that human consciousness is an unsoluble "world-riddle," a transcendent phenomenon in essential antithesis to all other natural phenomena—continues to be upheld in numerous writings. It is chiefly on this that the dualistic view of the world founds its assertion, that man is an altogether peculiar being, and that his personal soul is immortal; and this is the reason why the "Leipsic ignorabimus-speech" of Du Bois-Reymond has for twenty years been prized as a defence by all representatives of the mythological view of the world, and extolled as a refutation of "monistic dogma." The closing word of the discourse, "ignorabimus," was translated as a present, and this "ignoramus" taken to mean that "we know nothing at all"; or, even worse, that "we can never come to clearness about anything, and any further talk about the matter is idle." The famous "ignorabimus" address remains certainly an important rhetorical work of art; it is a "beautiful sermon," characterized by its highly-finished form and its surprising variety of philosophico-scientific pictures. It is well known, however, that the majority (and especially women) judge a "beautiful sermon" not according to the value of the thoughts embodied in it, but according to its excellence as an aesthetical entertainment. While Du Bois treats his audience at great length to disquisitions on the wondrous performances of the genius of Laplace, he afterwards glides over, the most important part of his subject in eleven short lines, and makes not the slightest further attempt to solve the main question he has to deal with—as to whether the world is really "doubly incomprehensible." For my own part, on the contrary, I have already repeatedly sought to show that the two limits to our knowledge of nature are one and the same; the fact of consciousness and the relation of consciousness to the brain are to us not less, but neither are they more, puzzling, than the fact of seeing and hearing, than the fact of gravitation, than the connection between matter and energy. Compare my discourse on *Freedom in Science and Teaching* (1878), pp. 78, 82, etc.]

17:

[*Immortality*. Perhaps in no ecclesiastical article of faith is the gross materialistic conception of Christian dogma so evident as in the cherished doctrine of personal immortality, and that of "the resurrection of the body," associated with it. As to this, Savage, in his excellent work on *Religion in the Light of the Darwinian Doctrine*, has well remarked: "One of the standing accusations of the Church against science is that it is materialistic. On this I would like to point out, in passing, that the whole Church-conception concerning a future life has always been, and still is, the purest materialism. It is represented that the material body is to rise again, and inhabit a material heaven." Compare also Ludwig Buchner, *Das zunkünftige Leben und die moderne Wissenschaft* (Leipsic, 1889); Lester Ward, "Causes of Belief in Immortality" (*The Forum*, vol. VIII., September 1889); and Paul Carus, *The Soul of Man: an Investigation of the Facts of Physiological and Experimental Psychology* (Chicago, 1891). Carus aptly points out the analogy between the ancient and the modern ideas with respect to light, and with respect to the soul. Just as formerly the luminous flame was explained by means of a special fiery matter (*phlogiston*), so the thinking soul was explained by the hypothesis of a peculiar gaseous soul-substance. We now know that the light of the flame is a sum of electric vibrations of the ether, and the soul a sum of plasma-movements in the ganglion-cells. As compared with this scientific conception, the doctrine of immortality of scholastic psychology has about the same value as the materialistic conceptions of the Red Indian about a future life in Schiller's "Nadowessian Death-Song."]

18:

[*Monistic Ethic*. All Ethic, the theoretical as well as the practical doctrine of morals, as a "science of law" (*Normwissenschaft*), stands in immediate connection with the view that is taken of the world (*Weltanschauung*), and consequently with religion. This position I regard as exceedingly important, and have recently upheld in a paper on "Ethik und Weltanschauung," in opposition to the "Society for Ethical Culture" lately founded in Berlin, which would teach and promote ethics without reference to any view of the world or to religion. (Compare the new weekly journal, *Die Zukunft*, edited by Maximilian Harden, Berlin, 1892, Nos. V.-VII.). Just as I take the monistic to be the only rational basis for all science, I claim the same also for ethics. On this subject compare especially the ethical writings of Herbert Spencer and those of B. von Carneri—*Sittlichkeit und Darwinismus* (1871); *Entwickelung und Glückseligkeit* (1886); and more particularly, the latest of all, *Der moderne Mensch* (Bonn, 1891); further, Wilhelm Streeker, *Welt und Menschheit* (Leipsic, 1892); Harald Höffding, *Die Grundlage der humanen Ethik* (Bonn, 1880); and the recent large work of Wilhelm Wundt, *Ethik, eine Untersuchung der Thatsachen und Gesetze des sittlichen Lebens* (Stuttgart, 2nd ed., 1892).]

19:

[*Homotheism*. Under the term homotheism (or anthropomorphism) we include all the various forms of religious belief which ascribe to a personal God purely human characteristics. However variously these anthropomorphic ideas may have shaped themselves in dualistic and pluralistic religions, all in common retain the unworthy conception that God (*Theos*) and man (*homo*) are organized similarly and according to the same type (homotype). In the region of poetry such personifications are both pleasing and legitimate. In the region of science they are quite inadmissible; they are doubly objectionable now that we know that only in late Tertiary times was man developed from pithecoid mammals. Every religious dogma which represents God as a "spirit" in human form, degrades Him to a "gaseous vertebrate" (*General Morphology*, 1866; Chap, xxx., God in Nature). The expression "homotheism" is ambiguous and etymologically objectionable, but more practical than the cumbersome word "Anthropotheism."]

20:

[*Monistic Religion*. Amongst the many attempts which have been made in the course of the last twenty years to reform religion in a monistic direction on the basis of advanced knowledge of nature, by far the most important is the epoch-making work of David Friedrich Strauss, entitled *The Old Faith and the New: A Confession* (11th ed., Bonn, 1881: *Collected Writings*, 1878). Compare M. J. Savage, *Religion in the Light of the Darwinian Doctrine*; John William Draper, *History of the Conflict between Religion and Science*; Carl Friedrich Retzer, *Die naturwissenschaftliche Weltanschauung und ihre Ideale, ein Ersatz fuer das religiöse Dogma* (Leipsic, 1890); E. Koch, *Natur und Menschengeist im Lichte der Entwickelungslehre* (Berlin, 1891). For the phylogeny of religion see the interesting work of U. Van Ende, *Histoire Naturelle de la Croyance* (Paris, 1887).]

21:

[*Freedom in Teaching*. The jubilee of the "Naturforschende Gesellschaft des Osterlandes" was celebrated in Altenburg on October 9, 1892, contemporaneously with the commencement of the brilliant celebration of the golden wedding of the Grand Duke and Duchess in Weimar. As exceptional as the celebration are the characteristics which distinguish this august couple. The Grand Duke Carl Alexander has, during a prosperous reign of forty years, constantly shown himself an illustrious patron of science and art; as Rector Magnificentißimus of our Thüringian university of Jena, he has always afforded his protection to its most sacred palladium—the right of the free investigation and teaching of truth. The Grand Duchess Sophie, the heiress and guardian of the Goethe archives, has in Weimar prepared a fitting home for that precious legacy of our most brilliant literary period, and has anew made accessible to the German nation the ideal treasures of thought of her greatest intellectual hero. The history of culture will never forget the service which the princely couple have thereby rendered to the human mind in its higher development, and at the same time to true religion.]

Made in United States
Orlando, FL
04 November 2026

90203411R00062